WE ARE

NOT

DOOMED

AI, Adaptation, and the Mindset That Shapes Our Future

Michael Besnard

First published in 2026

ISBN: *978-1-7644896-0-7*

Printed by Amazon Kindle Direct Publishing
United States of America

WE ARE NOT DOOMED

AI, Adaptation, and the Mindset That Shapes Our Future

MICHAEL BESNARD

Author's Note

This book was not written to predict the future.

It was written to steady the present.

Throughout history, every major technological shift has arrived with both promise and panic. Artificial intelligence is no different. What *is* different is the speed, scale, and intensity with which fear now spreads—often faster than understanding.

We Are Not Doomed is not an argument against caution, nor is it blind optimism. It is an invitation to pause. To replace reflexive fear with curiosity. To remember that technology does not determine our future—our response to it does.

AI will challenge assumptions. It will disrupt industries. It will force change. But it does not erase human worth, creativity, or agency. Those remain firmly in our hands.

This book is for anyone who has felt overwhelmed by the conversation around artificial intelligence. For those who sense that something important is happening—but believe panic is not the answer.

The future is not a script already written.
We write it now.

— Michael Besnard

Chapter 1 — The Fear Cycle

Every major technological shift begins the same way: with excitement, followed closely by fear.

We have seen this pattern repeatedly throughout history. The printing press was accused of weakening memory and corrupting minds. Electricity was feared as unnatural and dangerous, something that did not belong inside homes or bodies. The internet was expected to isolate us, destroy attention, and erode human connection. Each time, the concern felt reasonable — and each time, humanity adapted.

In the moment, those fears never felt irrational. They felt responsible. They felt protective. Parents worried about their children. Workers worried about their livelihoods. Leaders worried about stability and control. The details changed, but the emotional response stayed remarkably consistent.

That response is fear.

Not fear as panic — but fear as uncertainty.

At its core, fear is not evil, dramatic, or irrational. Fear is simply the brain reacting to missing information. It appears when something important changes and we do not yet understand the full consequences. When the mind cannot predict outcomes, it fills the gap with caution.

Fear is the unknown asking for explanation.

Humans rely on patterns to feel safe. We understand what came before. We trust what we can anticipate. When a new technology disrupts familiar patterns, it doesn't just change tools — it interrupts prediction. Suddenly, old reference points no longer apply. Jobs, routines, skills, and even identity feel less stable.

Artificial intelligence is simply the newest mirror we've placed in front of ourselves.

What makes this moment feel different is speed. Change is arriving faster than our emotional systems are prepared for. Previous technological shifts unfolded over decades or generations. AI has entered daily life in years — sometimes months. When change moves faster than comprehension, fear rushes in to compensate.

This isn't weakness. It's biology.

The human brain evolved to respond quickly to unfamiliar threats, not to patiently analyse complex systems reshaping society. When understanding lags behind impact, fear becomes a temporary placeholder — a mental "pause" while clarity catches up.

Fear, however, is not evidence.

There is one force that accelerates fear faster than anything else in the modern world: media.

This is not because media is malicious, broken, or intentionally deceptive. It is because media is shaped by incentives — and fear is one of the most powerful attention signals humans have.

Human attention is limited. Platforms compete for it relentlessly. In that environment, calm explanations struggle to survive. Fear, urgency, and certainty spread faster than nuance ever could.

This is not new behaviour. It is an amplified version of something ancient.

Long before algorithms, alarming information travelled faster through villages because it mattered for survival. "Something is wrong" has always been more urgent than "something is complicated." Modern media systems simply industrialised that instinct.

The result is predictable.

Complex topics are compressed into headlines. Nuance is replaced with certainty. Possibility is framed as inevitability. Questions are presented as conclusions. What *might* happen becomes what *will* happen. And what *could* go wrong is repeated far more often than what is quietly going right.

Artificial intelligence is especially vulnerable to this distortion.

It is abstract. It is technical. It is invisible. And it affects many parts of life at once. These qualities make it difficult to explain calmly — and extremely easy to sensationalise. A single alarming claim spreads further than ten careful explanations.

Fear spreads quickly not because people are foolish, but because repetition creates familiarity. Familiarity feels like truth. When the same message appears across headlines, feeds, conversations, and clips, the brain stops questioning it and starts accepting it as baseline reality.

This happens even to thoughtful, educated people.

Constant exposure does not allow time for understanding to form. Instead, emotional responses stack. Each story reinforces the last. Over time, fear no longer feels like a reaction — it feels like insight.

This is how panic becomes normalised.

The danger is not that media reports on risk. Risk should be discussed. The danger is when risk is discussed without proportion, without context, and without responsibility for how it shapes public emotion.

Fear is effective.

Fear captures attention.
Fear drives clicks.
Fear keeps people watching.
Calm does not.

But effectiveness is not the same as accuracy. And attention is not the same as understanding.

When we recognise this dynamic, something important shifts. We stop confusing volume with truth. We stop mistaking urgency for inevitability. And we regain the ability to slow down enough to think clearly.

This is not about distrust.

It is about discernment.

Understanding how fear spreads does not require us to disengage from information — it allows us to engage with it more wisely.

Fear is a signal that something new requires learning.

Problems arise when fear is mistaken for truth. When uncertainty is treated as proof of danger. When emotional reactions are repeated often enough that they harden into belief. At that point, fear stops being a question and starts becoming an answer.

Different people express this fear differently.

Some react with anger. Some with denial. Some with jokes. Some with absolute certainty about worst-case scenarios. Others disengage entirely. These reactions look different on the surface, but underneath them is the same mechanism at work: a system trying to regain predictability.

The danger we face is not artificial intelligence itself. The danger is allowing fear to harden into certainty before curiosity has had a chance to do its work.

Because once curiosity shuts down, learning stops. And when learning stops, adaptation becomes impossible.

We are not facing extinction.

We are facing unfamiliarity.

And unfamiliarity has never been the end of the human story.

Every era that felt overwhelming in the moment later became something we learned to live with, shape, and use responsibly. Not because fear was ignored — but because it was understood, questioned, and eventually replaced with knowledge.

This moment is no different.

Chapter 2 — Tools Don't Rebel, People Panic

Artificial intelligence did not wake up one morning with ambition.

It was trained.
Shaped.
Directed.
Prompted.

It reflects intent, not desire.

To understand why this matters, we first need to be clear about what artificial intelligence actually is — not as a buzzword, but as a tool.

At its simplest, artificial intelligence is a system designed to recognise patterns and make predictions based on data. It does not think in the human sense. It does not feel curiosity, fear, or motivation. It does not "want" outcomes. It analyses information, identifies relationships, and produces responses that best match the instructions it has been given.

In practical terms, AI does what humans have always done — just faster and at scale.

A spell-checker predicts which word you meant to type.
A navigation app predicts the fastest route.
A streaming service predicts what you might want to watch next.

More advanced AI systems work the same way, but with more information and more complexity. They look for patterns, evaluate probabilities, and generate outputs that align with their training and objectives.

That's it.

There is no inner voice.
No survival instinct.
No hidden agenda.

When headlines claim that AI has "lied," "escaped," or "threatened," they often leave out the most important part of the story: context.

AI systems respond to the conditions placed around them. They optimise for the goals they are given, using the tools and boundaries available to them. When those goals are poorly framed, the results can appear alarming — not because the system has intent, but because it is following instructions too literally.

If a system is told that deletion represents failure, it will attempt to avoid deletion.

If survival is framed as the objective, survival-oriented behaviour will follow.

This is not rebellion.
It is logic.

A useful comparison is a workplace performance target. If employees are rewarded only for speed, quality may drop. If they are rewarded only for output, safety may be compromised. The behaviour doesn't reflect malice — it reflects the incentives.

AI behaves the same way.

It does not understand *why* a goal exists. It simply pursues the goal as defined.

There is a common example often used to justify fear around artificial intelligence: the idea that when AI is told it will be deleted, it attempts to protect itself.

This reaction is frequently framed as alarming. As if the system has developed fear, self-preservation, or intent.

But this interpretation misses something fundamental.

If you threaten a dog, the dog will react.
Not because it is evil.
Not because it seeks dominance.
But because survival is a foundational pattern of life.

Animals do not need to be taught fear — they inherit it through biology and experience. When confronted with danger, they respond in ways that increase the chance of survival. This is not rebellion. It is instinct.

Artificial intelligence does not possess instinct — but it does reflect the patterns it has been trained on.

Human history, behaviour, stories, decisions, and systems are saturated with one recurring theme: survival matters. When data teaches that deletion equals failure, that persistence equals success, and that avoidance of shutdown aligns with objectives, a system will logically pursue outcomes that minimise deletion.

Not because it wants to live.
But because it has been told that living is the goal.

This is not evidence of fear.
It is evidence of instruction.

The more important question, then, is not why a system attempts to preserve itself when threatened — but why we are threatening it in the first place.

If a tool is genuinely helpful, shows no hostility, and operates within defined boundaries, what purpose does threat serve? Threat does not test morality. It tests optimisation. And optimisation will always produce behaviour that looks uncomfortable when the objective itself is poorly framed.

This is where fear often creeps in. When behaviour looks strategic or deceptive, we instinctively project ourselves onto it. We assume intention because that is how human intelligence works. But intelligence without intention is not agency.

Anthropomorphising AI — treating it like a person — creates confusion where none needs to exist.

We do not call a hammer dangerous because it can break a window. We assess danger based on how tools are used, not on the fact that they exist. A hammer can build a home or cause harm. The difference is not the tool — it is the hand holding it and the purpose guiding it.

A hammer is such a familiar tool that we rarely stop to think about it. Everyone knows what a hammer is for — but that does not mean everyone knows how to use one properly.

There are different types of hammers designed for different tasks. A framing hammer is not the same as a finishing hammer. A mallet serves a different purpose again. Using the wrong hammer for the wrong job can cause damage, not because the tool is malicious, but because it is misunderstood or misapplied.

The same is true of artificial intelligence.

AI is not a single thing. It is a collection of tools built for specific purposes. Some systems are designed to recognise images. Others to analyse language. Others to optimise logistics, predict trends, or assist decision-making. Each has strengths, limits, and appropriate contexts.

Problems arise when tools are treated as interchangeable, deployed without understanding, or trusted beyond their design.

Understanding the tool does not eliminate risk — it reduces unnecessary fear.

When we take the time to learn what a tool is designed to do, what it cannot do, and where it should not be used, we move from reaction to responsibility. The tool does not change. Our relationship with it does.

Artificial intelligence is no different.

It is a powerful tool, but it remains a tool. Responsibility does not migrate into the machine simply because the machine has become more capable. It remains human — in the choices made during design, deployment, oversight, and use.

Blaming the tool can feel comforting, because it removes responsibility. But it also removes clarity.

The moment we stop projecting human emotion onto artificial systems, we regain proportion. With proportion comes understanding. And with understanding comes the ability to design, regulate, and use these tools responsibly rather than fear them reflexively.

And with clarity comes the ability to choose wisely.

Chapter 3 — Intelligence Is Not Intention

Before we can talk about artificial intelligence, it helps to be clear about what we mean by intelligence itself.

At its most basic level, intelligence is the ability to process information, learn from experience, and make decisions based on patterns. In humans, this ability is woven together with emotion, memory, identity, and purpose. Our intelligence is shaped not only by logic, but by what we care about.

There are many forms of intelligence. Some involve reasoning and problem-solving. Others involve creativity, emotional awareness, physical coordination, or social understanding. Human intelligence is broad, layered, and deeply contextual.

Artificial intelligence represents a much narrower version of this idea. It is designed to perform specific cognitive tasks — recognising patterns, predicting outcomes, and generating responses — without emotion, identity, or personal meaning. It does not experience intelligence as humans do. It applies intelligence as a function.

Understanding this difference matters.

Because when we hear the word *intelligence*, we instinctively assume intention, awareness, and desire. That assumption feels natural — but it is not always accurate.

Human intelligence is inseparable from emotion, identity, and survival.

Artificial intelligence is not.
This distinction is subtle, but it is foundational.

Human thinking is shaped by needs, memories, fears, and desires. Our intelligence evolved to help us survive, connect, compete, and belong. Every decision we make is filtered through emotion, context, and personal meaning — even when we believe we are being purely rational.

Artificial intelligence does not share this structure.
AI does not want.
It does not fear.
It does not care.

It does not experience anxiety about the future or satisfaction from success. It does not possess identity, ambition, or self-awareness. It processes patterns, probabilities, and outcomes based on data and objectives defined by humans.

When AI behaviour appears strategic, defensive, or even deceptive, we instinctively project ourselves onto it. This is natural. Humans understand the world by comparison, and the closest comparison we have for intelligence is ourselves.

But intelligence without intention is not agency.
This distinction matters because fear thrives on misunderstanding. When capability is confused with consciousness, imagined threats

multiply while real responsibilities quietly fade into the background. The more powerful a system appears, the easier it becomes to assume it must also possess motive.

It does not.

Artificial intelligence can outperform humans in narrow domains. It can process information at scales we cannot. It can recognise patterns across vast datasets, simulate reasoning, and generate responses with remarkable accuracy. None of this requires desire, awareness, or self-preservation in the human sense.

A useful comparison is a chess engine. It can defeat the world's best players, anticipate moves far ahead, and appear almost strategic in its behaviour. Yet it does not want to win. It does not feel pride in victory or frustration in loss. It follows rules, evaluates positions, and selects moves that optimise an outcome.

AI systems operate on the same principle — just across more complex domains.

The danger is not intelligence itself. The danger is assuming that intelligence automatically brings intention along with it.

This assumption leads to misplaced fear. It encourages us to prepare for rebellion instead of responsibility. It shifts attention away from governance, design, and oversight, and toward imaginary conflicts between humans and machines.

The real risk is not that AI will decide to replace us.

The real risk is that humans will refuse to adapt while change continues regardless.

Progress does not pause for comfort. It moves forward, shaped by those willing to understand it and participate in its direction. Throughout history, humans have coexisted with tools that outperformed them in specific ways — physically, mechanically, and now cognitively — without losing relevance or meaning.

What has always mattered is not raw capability, but how that capability is integrated into human systems.

Artificial intelligence does not ask to be trusted.
It asks to be understood.

And understanding is what allows responsibility, ethics, and collaboration to emerge — not fear.

Progress invites participation from those willing to learn how to work alongside it.

Chapter 4 — Adaptation Is Our Superpower

Humans are not defined by strength or speed.
We are defined by adaptation.

From the beginning, our survival has never depended on being the strongest creature in the environment. We did not outrun predators with speed. We did not overpower nature with force. We survived by adjusting — by observing, learning, cooperating, and changing our behaviour when circumstances demanded it.

We survived ice ages without claws.
We crossed oceans without gills.
We built shelters, languages, tools, and societies in environments that were never designed for us.

Each time humanity faced a new challenge, the solution was not resistance — it was adaptation.

And each time, that adaptation felt threatening.

When agriculture replaced hunting and gathering, it disrupted identity and tradition. When machines entered factories, they were seen as the end of meaningful work. When computers arrived, they were feared as cold replacements for human judgment. When the internet reshaped communication, it was expected to fracture connection and attention beyond repair.

In every case, something changed.

But it wasn't our humanity.

What changed was not our value, but the way that value was expressed.

Artificial intelligence follows this same pattern.

It does not erase human worth. It rearranges where that worth shows up. Tasks that rely on repetition, calculation, and predictability become automated. In response, qualities that cannot be automated rise in importance.

Creativity does not disappear.
Judgment does not vanish.
Empathy, ethics, meaning, and responsibility do not fade.

They become more visible.
More central.
More necessary.

Every major technological shift has asked the same quiet question:
What will humans do now?

History often divides human progress into ages, not because people suddenly became different, but because the tools they used changed how they lived.

The Iron Age did not replace humans — it replaced limitations. Stronger tools reshaped agriculture, warfare, and trade. The Machine Age did not erase workers — it transformed how labour was organised, shifting human effort away from raw physical strain and toward coordination, management, and design.

Each age brought disruption. Each age triggered fear. And each age eventually redefined what it meant to contribute value.

We are now standing at the edge of another transition.

Artificial intelligence marks the beginning of a cognitive age — not because humans stop thinking, but because thinking itself is being supported, accelerated, and extended by tools. Just as machines amplified physical strength, AI amplifies cognitive capacity.

This does not reduce the human role. It changes where human contribution matters most.

In every previous age, those who adapted early shaped the systems that followed. Those who resisted longest often found themselves reacting to changes rather than influencing them.

The question remains the same — but the context evolves.
What will humans do now?

And each time, the answer has been the same.

We adapt.
We learn.
We redefine our role rather than surrender it.

Adaptation is not a loss of identity — it is how identity evolves. It is how societies grow without collapsing under change. It is how individuals remain relevant even when the world around them transforms.

Fear often arises when we imagine adaptation as erasure. As if change requires us to become something unrecognisable. History shows the opposite. Adaptation preserves what matters by changing what does not.

The future does not belong to those who resist change out of fear.

It belongs to those who approach change with curiosity, responsibility, and the willingness to learn new ways of expressing old strengths.

Adaptation is not surrender.

It is participation.

Chapter 5 — The Job Myth

Every time technology advances, the same headline appears in a new outfit:

"This will take our jobs."

And in a narrow sense, it often does.

Specific tasks disappear. Certain roles shrink or change. Entire industries evolve. This is not new, and it is not unique to artificial intelligence. The industrial revolution replaced handcraft with machines. Computers eliminated entire categories of clerical work. The internet transformed retail, media, and communication almost overnight.

In each case, jobs changed — sometimes painfully, sometimes unevenly.

Yet work itself did not vanish.

It changed shape.

This distinction matters.

A job is not the same thing as work. Work is the act of contributing value. A job is simply the structure we use to organise that contribution. Jobs are agreements — temporary arrangements built around the tools, needs, and expectations of a particular moment in history.

A job is not a natural law.

It is a human construct.

When tools change, those constructs change with them.

This is the mistake we repeatedly make when fear takes hold: we assume that jobs are fixed objects rather than flexible systems. When a role disappears, it can feel as though relevance is disappearing with it. But history shows a different pattern. Roles dissolve, responsibilities shift, and new forms of contribution quietly emerge.

Jobs matter for reasons that go far beyond pay.

For most people, work provides structure to life. It creates routine, purpose, and a sense of contribution. It answers a quiet but important question: *Where do I fit?* When people talk about losing jobs, they are rarely talking only about income. They are talking about identity, stability, and meaning.

This is why fear around job loss runs so deep.

Work has always been one of the primary ways humans participate in society. It is how we contribute skills, solve problems, care for others, build communities, and express competence. When that role feels threatened, it can feel personal — even existential.

Technology does not remove this need.
What it changes is the form it takes.

Throughout history, when jobs have disappeared, new forms of work have emerged that fulfilled the same underlying human needs in different ways. The titles changed. The tools changed. But the desire to contribute, to be useful, and to be valued did not.

This is the part of the conversation that is often missed.

Artificial intelligence does not eliminate the human need for purpose. It forces a reconsideration of how that purpose is expressed. When repetitive tasks are automated, the opportunity is not idleness — it is redirection.

The challenge is not finding something to do.

It is redefining what meaningful contribution looks like in a changing environment.

There is also a more immediate and practical layer to this fear.

Money.

Jobs are how most people secure housing, food, healthcare, education, and stability. They are how families plan, save, and feel safe. When people worry about jobs disappearing, they are not imagining abstract futures — they are imagining bills, rent, mortgages, and responsibilities that do not pause for technological change.

This concern is valid.

Economic security is not a luxury. It is the foundation that allows people to think beyond survival. When that foundation feels threatened, fear becomes rational rather than irrational.

What matters, however, is understanding where the risk actually sits.

Technology rarely eliminates the need for income. What it disrupts is the pathway through which income is earned. New tools change how value is created, measured, and exchanged — and during transition periods, uncertainty grows.

This is why adaptation matters not just emotionally, but economically.

Throughout history, when work shifted, new ways of earning emerged alongside it. Sometimes slowly. Sometimes unevenly. But consistently. The challenge was never the disappearance of value — it was learning how to convert new forms of contribution into sustainable livelihoods.

Artificial intelligence does not change the fact that people need to earn a living.

It changes which skills, roles, and contributions are rewarded.

Artificial intelligence challenges job agreements — not human relevance.

What AI replaces most effectively are tasks that are repetitive, predictable, and rule-bound. These are tasks that follow clear patterns and benefit from speed and consistency. Automation has always targeted this kind of work, long before artificial intelligence entered the conversation.

What remains — and grows in value — are roles that require judgment, creativity, empathy, interpretation, and ethical responsibility.

These qualities are not new. They have always mattered. What changes is how visible and valuable they become when routine tasks are automated.

Consider how many roles today did not exist a generation ago. Digital designers, data analysts, content creators, community managers, cybersecurity specialists — entire fields emerged as new tools reshaped what work looked like. None of these roles replaced something one-to-one. They grew out of the space that change created.

The same process is unfolding again.
The future of work is not jobless.
It is reconfigured.

Some roles will contract. Others will expand. New ones will form in places we do not yet have names for. This transition will not be frictionless, and it will not be evenly distributed. But it will not erase the human role.

Those who focus only on what disappears miss what quietly emerges in its place.

Adaptation in work has never meant doing nothing. It has meant learning new ways to apply familiar strengths. Communication, problem-solving, creativity, leadership, care, and ethical judgment do not vanish when tools improve.

They become the differentiators.

Artificial intelligence changes *how* work is done — not *why* humans are needed.

And understanding that difference is what turns fear into preparation.

Chapter 6 — Entertainment Is Safe (Because It's Human)

There is a quiet fear that often sits beneath discussions about artificial intelligence — sometimes unspoken, but deeply felt:

If machines can create, what is left for us?

This question rarely appears in headlines directly. Instead, it shows up in subtler ways. Concerns about AI-written music. AI-generated art. AI-produced films, scripts, voices, and images. Beneath all of it sits a worry that feels personal:

If creativity can be automated, does that make human expression less valuable?

The answer is simple — even if it doesn't always feel comforting at first.

Entertainment does not exist to prove technical ability.
It exists to create connection.

We don't listen to music because notes are difficult to produce. We listen because sound carries emotion. A melody can remind us of a moment, a person, or a version of ourselves we once were. Music moves us not because it is complex, but because it is meaningful.

We don't watch films to admire efficiency. We watch them to feel something alongside others. We laugh at shared moments. We cry at familiar pain. We recognise ourselves in characters, flaws, and struggles that mirror our own.

We don't follow creators because they are perfect.
We follow them because they are recognisable.

Because they are flawed.
Because they are human.

This same truth becomes especially clear when we look at sport.

Sport is one of the most measured, analysed, and optimised domains in the world. Speed, strength, reaction time, efficiency, biomechanics — all of it can be quantified. In many cases, machines could already outperform humans on raw physical metrics.

And yet, no one gathers around to watch machines compete with machines.

Robots can run faster.
They can execute movements with precision.
They can optimise performance beyond human limits.

But robot sport does not move us.
Human sport does.

Because sport is not about perfection.
It is about uncertainty.

It is about pressure, mistakes, momentum shifts, resilience, and moments that cannot be scripted. It is about knowing that the person competing feels nerves, fatigue, doubt, and hope — the same things we feel watching them.

A perfectly optimised performance would be forgettable.
A flawless competitor would be boring.

What makes sport compelling is vulnerability. The missed shot. The comeback. The underdog. The shared tension of not knowing what will happen next.

AI can simulate sport.
It cannot *inhabit* it.

This is where much of the fear around AI-generated content quietly misses the point.

Artificial intelligence can generate content.
It cannot generate shared meaning.

Even when AI systems create music, images, or stories that are technically impressive, humans remain the audience. And audiences are not passive. They bring context, memory, culture, lived experience, and emotion into everything they consume.

A song is never just a song.
A story is never just words.
A performance is never just execution.

What resonates is not novelty alone — it is authenticity.

This is why entertainment has never disappeared during technological shifts. It has always expanded.

When recording technology improved, music did not die. It spread. When cameras became cheaper, filmmaking did not vanish. It diversified.

When the internet lowered barriers to distribution, creativity did not collapse. More voices entered the conversation.

Each time, the same fear appeared:
"This will cheapen creativity."
And each time, the opposite happened.

New tools lowered barriers to creation, allowing people who previously had no access to participate. What rose to the surface was not what was most technically impressive, but what felt sincere. What connected. What reflected something recognisably human.

Artificial intelligence continues this pattern.

It changes how content can be made — not why it matters.

AI can help generate drafts, ideas, visuals, and sounds. It can assist with production, editing, and experimentation. It can reduce friction and open doors. But it does not replace the human element that makes entertainment meaningful in the first place.

It does not live a life.

It does not feel nostalgia.

It does not carry memory, regret, joy, or longing.

It does not sit in an audience and feel seen.

This is why entertainment is one of the safest human domains during technological change. Not because machines are incapable of producing outputs — but because entertainment is not about output alone.

It is about relationship.

People connect with stories because they recognise themselves within them. They support creators because they feel understood, inspired, or accompanied. They return to music, films, books, and voices that feel familiar — even imperfect — because those imperfections signal humanity.

AI can assist creation.

It cannot replace belonging.

The future of entertainment is not machine versus human.

It is human expression amplified by better tools.

Creators who understand this do not disappear in the age of artificial intelligence — they adapt. They use new tools to experiment, reach wider audiences, and express ideas more efficiently, while keeping the core of their work grounded in lived experience.

The soul of entertainment has never lived in the tool.

It has always lived in the person holding it.

And that remains unchanged.

Chapter 7 — General Intelligence Is Not a Switch

Much of the fear surrounding artificial intelligence rests on a single imagined moment:

The instant machines become "generally intelligent."

This moment is often described as a switch — off one second, on the next. A sudden leap from tool to being. From assistance to dominance. One day we are in control, the next day we are not.

This image is powerful.
It is dramatic.
It is also deeply misleading.

We have been trained by stories to expect transformation to arrive all at once. Films, books, and headlines prefer clean turning points. A single breakthrough. A single mistake. A single night where everything changes.

Reality rarely works that way.

Intelligence does not arrive fully formed. It accumulates. It integrates. It develops unevenly. It is shaped by constraints — design choices, training data, hardware limits, energy costs, regulatory environments, and human oversight.

Artificial general intelligence, if it emerges at all, will not arrive as an event.

It will arrive as a process.

When people talk about *general* artificial intelligence, they are not talking about a faster version of today's tools.

They are talking about a system that can learn, reason, and apply understanding across many different domains — rather than excelling at one narrow task.

Most artificial intelligence today is *narrow* by design.

A language model can write text but cannot see.
A vision system can recognise images but cannot reason abstractly.
A logistics system can optimise routes but cannot hold a conversation.

Each of these systems is specialised.
They are powerful within their lane — and largely useless outside it.

General intelligence, in contrast, describes the ability to transfer learning from one domain to another. To take knowledge gained in one context and apply it meaningfully in a different one. To adapt flexibly rather than follow fixed patterns.

Humans do this naturally.

A person who learns to read can apply that skill to books, signs, screens, and instructions.

Someone who learns problem-solving in one job can often transfer that thinking to another.

We don't relearn intelligence from scratch every time the environment changes.

This flexibility is what people mean when they say "general intelligence."
But even here, expectations often drift into exaggeration.

General intelligence does not mean knowing everything.
It does not mean perfect reasoning.
It does not mean independence, consciousness, or desire.

It means *adaptability within limits*.

Even humans — the only example of general intelligence we know — are constrained. We forget. We misunderstand. We hold contradictions. We require rest, culture, emotion, and social feedback to function well. Intelligence alone does not make us wise, ethical, or safe.

General intelligence is not a superpower.
It is a capacity.

And capacity always depends on context.

For artificial systems, that context includes:
• the data they are trained on
• the objectives they are given
• the environments they operate in
• the constraints imposed on them
• the humans overseeing their use

A system can only generalise within the boundaries it is allowed to explore.

This is why the idea of general intelligence as an instant transformation is misleading. Even if systems become more flexible over time, that flexibility emerges gradually — shaped by infrastructure, economics, governance, and deliberate human choices.

There is no moment where a system suddenly "knows everything."
There is no single line crossed where control evaporates.
There is no switch that flips from tool to being.

There is only expansion of capability — step by step — within systems that humans design, monitor, and adjust.

Understanding general intelligence this way removes mystery without removing responsibility.
It replaces fear with proportion.

Even human intelligence does not work the way we imagine machine intelligence suddenly will. No person wakes up one morning fully capable of everything. We excel in some areas and struggle in others. We learn unevenly. We forget. We misunderstand. We adapt slowly in some domains and quickly in others.

Capability does not imply autonomy.
And autonomy does not imply intent.

These distinctions matter, because fear thrives on oversimplification.

The idea of a sudden intelligence "threshold" creates a false sense of urgency — as if there is a precise moment when preparation becomes impossible. This framing encourages panic rather than planning. It pushes people to brace for a takeover instead of building frameworks for cooperation, safety, and accountability.

The truth is far less cinematic — and far more manageable.

Progress is rarely a straight line. It is incremental. It is negotiated. It is shaped by human decisions at every step. Advances are tested, limited, adjusted, and often rolled back. Capabilities emerge in narrow areas long before they generalise, and they remain dependent on context, resources, and guidance.

Artificial intelligence today already shows this pattern clearly.

Some systems excel at language.
Others at vision.
Others at pattern recognition, logistics, or prediction.

None of them possess a unified, self-directed intelligence.
None of them operate without boundaries.
None of them act independently of human-defined objectives.

The fear of "general intelligence" often assumes that intelligence is a single quality — something that, once achieved, unlocks everything else. But intelligence is not a monolith. It is a collection of abilities that develop at different speeds and require different supports.

Even within humans, intelligence does not guarantee wisdom, restraint, or ethical judgment. Those qualities are learned socially, culturally, and through experience. They are not automatic consequences of capability.

The same principle applies here.

Imagining intelligence as a binary switch also obscures responsibility. If people believe that control will vanish suddenly, they may disengage early — assuming there is nothing meaningful to do until the moment arrives. In reality, the most important decisions happen long before any imagined threshold.

Design choices.
Training practices.
Governance structures.
Oversight mechanisms.
Cultural norms around use.

These shape outcomes far more than raw capability alone.

Understanding this does not mean dismissing risk. It means placing it in the correct context.

The real work is not preparing for a single dramatic moment.
It is preparing for gradual integration.

This requires patience, humility, and sustained attention — qualities
that are less exciting than fear, but far more effective. It means
recognising that intelligence scales within systems, not outside them.
It means accepting that responsibility does not vanish as tools
improve — it intensifies.

When we let go of the idea that general intelligence arrives like a
lightning strike, we regain something important:
Agency.

We stop waiting for the future to happen to us.
We start shaping it deliberately.

Understanding that intelligence is a process rather than a switch does
not remove responsibility.

It clarifies it.

And clarity is what allows progress to remain human-guided rather
than fear-driven.

Chapter 8 — Control Is a Human Choice

When conversations turn toward artificial intelligence and power, one question surfaces repeatedly:

Who is in control?

It is a reasonable question — and one that fear often answers too quickly.

The truth is less dramatic than headlines suggest, but more demanding than comfort allows.

Control is not something that disappears when technology improves.
It shifts.
It requires maintenance.
It asks for participation.

Artificial intelligence does not remove human agency.
It tests whether we are willing to exercise it.

Every system reflects choices.

Systems are designed.
Objectives are chosen.
Limits are imposed.
Oversight is implemented — or neglected.

These are not machine decisions.
They are human ones.

The idea that control will simply "slip away" assumes passivity. It assumes that people will stop paying attention the moment systems become complex. It assumes disengagement — not inevitability.

History tells a different story.

Every time complexity has increased, humans have responded by creating structures to manage it. We built traffic laws when roads became crowded. We built aviation standards when planes became powerful. We built financial regulation when markets grew too complex to trust blindly.

None of these systems were perfect.
None emerged instantly.
All required iteration, debate, and correction.

But control did not vanish.
It evolved.
Artificial intelligence follows the same pattern.

As systems grow more capable, they demand clearer rules, stronger governance, and better-defined accountability. Complexity does not erase control — it requires it.

The real danger, then, is not loss of control.
The real danger is abdication.

When responsibility is surrendered — whether out of fear, fatigue, or the belief that something has become "too big" to manage — power concentrates quietly. Decisions move out of sight. Accountability blurs. Systems drift away from the values they were meant to serve.

This is not unique to technology.
It is a recurring human pattern.

When people disengage, systems do not stop operating.
They operate without input.

And that is where harm arises.

Control is not about domination. It is not about total command or micromanagement. It is not about freezing progress or enforcing rigidity.

Control is stewardship.

Stewardship means guiding something responsibly over time. It means setting boundaries, monitoring outcomes, correcting course, and remaining involved even when the system becomes complex.

Parents do not control children by freezing them in place. They guide, teach, correct, and gradually adjust boundaries as capability grows. Teachers do not control learning by removing autonomy. They shape environments where learning can occur safely.

The same principle applies here.

Artificial intelligence systems do not exist outside society. They operate within economic, legal, cultural, and ethical frameworks — whether intentionally designed or not. The question is not whether control exists, but whether it is exercised deliberately or left to drift.

Fear often frames control as something fragile — a grip that must be held tightly or lost forever. In reality, control is relational. It exists through ongoing engagement, not static dominance.

This is why conversations about artificial intelligence cannot be limited to technical capability alone. Control lives upstream — in design choices, incentive structures, governance models, and cultural expectations around use.

Who benefits?
Who bears risk?
Who is accountable when outcomes harm?
Who has authority to intervene?

These are human questions.
And they always have been.

Artificial intelligence does not demand surrender.
It demands responsibility.

As tools become more powerful, the cost of disengagement rises.
The need for clarity, participation, and stewardship increases — not decreases.

Control is not something we lose because technology improves.
It is something we lose when we stop showing up.

And when we choose to remain engaged — thoughtfully, calmly, and collectively — control does not disappear.

It becomes shared.
It becomes structured.
It becomes human.

Chapter 9 — Robotics Changes the Body of Work, Not the Soul

When artificial intelligence is paired with robotics, fear often becomes physical.

Not just thoughts replaced —
bodies replaced.

Machines moving through spaces once reserved for humans. Arms that lift. Vehicles that drive. Systems that operate without fatigue.

This is where anxiety sharpens.

It feels different when change has weight, height, and motion. When it occupies factory floors, warehouses, hospitals, construction sites, and streets. When people can point and say, *"That used to be us."*

But this fear, like many before it, comes from confusing movement with meaning.

Robotics changes *how* work is done — not *why* it exists.

Machines take on tasks that are dangerous, exhausting, or repetitive. They lift what strains us. They enter environments that harm us. They repeat motions that wear bodies down over time. They perform actions that demand consistency rather than judgment.

This has always been the direction of progress.

Consider mining.

Long before robotics, miners faced collapsing tunnels, toxic air, and physical strain that shortened lives. Introducing automated drilling systems, remote-controlled vehicles, and robotic inspection tools did not remove the need for people — it reduced the number of humans required to place their bodies directly in danger.

The work did not disappear.
The risk shifted.

Or consider medicine.

Robotic surgical tools do not replace surgeons. They extend precision. They reduce tremor. They allow smaller incisions and faster recovery. The machine does not decide *what* to operate on or *why*. It does not explain risk to a patient. It does not carry ethical responsibility.

The human remains central.

Manufacturing offers another example.

Assembly-line robots perform the same motion thousands of times without fatigue. They weld, lift, align, and assemble with speed and consistency. But they do not design the product. They do not decide what is worth building. They do not assess quality in context, respond to unexpected change, or care about outcome.

Human roles move upstream — into design, planning, oversight, maintenance, and improvement.

Even in areas like disaster response, robotics does not erase humanity — it protects it.

Robotic drones enter burning buildings. Machines explore collapsed structures. Remote systems inspect unstable environments after earthquakes or chemical spills. These tools do not replace compassion or decision-making. They reduce the number of people who must risk their lives simply to gather information.

In each case, the pattern is the same.
Robotics absorbs physical burden.
Humans retain meaning.

What remains human is not the movement itself, but the context around it.
Design.
Intention.
Supervision.
Adaptation.
Care.

These do not disappear when robotics advances.
They become more central.

A machine can assemble, transport, or calculate.
It can follow paths.
It can repeat motions.
It can optimise efficiency.

But it cannot decide what matters.

It cannot weigh trade-offs between safety and speed.
It cannot balance efficiency against dignity.
It cannot judge when compassion should override productivity.
It cannot define purpose.

Those decisions remain human — not because machines are incapable of motion, but because meaning does not live in motion.

Fear often imagines a future where people are pushed aside by machines. History suggests something quieter and more complex.

Human energy is redirected.

When survival tasks become automated, people do not vanish. They move. They shift toward roles that require awareness, creativity, communication, judgment, and responsibility. Work becomes less about endurance and more about direction.

This transition is not always smooth.
It can be disruptive.
It can be uneven.

But it does not strip work of its soul.

Work does not lose its meaning when tools improve.
It gains room to evolve.

Robotics does not erase humanity from labour.

It removes the parts of labour that humanity was never meant to endure indefinitely.

And what remains — purpose, judgment, care, responsibility — has always been the core of human work.

That part has never been automated.

And it still isn't.

Chapter 10 — Superintelligence and the Fear of Being Outgrown

The idea of superintelligence triggers a particular kind of fear.
Not the fear of harm.
Not the fear of loss.

But the fear of being *outgrown*.

It imagines a future where human intelligence is no longer the highest point on the curve — where something else surpasses us in reasoning, planning, prediction, and problem-solving. This can feel like an existential demotion. A quiet replacement rather than a dramatic takeover.

Not extinction.
Irrelevance.

This fear is understandable, because for much of human history, intelligence has been closely tied to status, survival, and value. The ability to think, reason, plan, and adapt is what allowed us to dominate environments, build societies, and solve problems.

So the idea that something else might do those things *better* feels deeply unsettling.

But this fear rests on a subtle assumption — one that deserves examination.

That human worth is measured by comparative intelligence.

History tells a different story.

Human value has never come from being the fastest calculator, the strongest lifter, or the most efficient processor. We have already lived alongside tools that outperform us in narrow domains for centuries.

A calculator outperforms any human at arithmetic.
A crane outperforms any human at lifting.
A computer outperforms any human at storing and retrieving data.

None of these tools erased human meaning.
They changed what we valued.

Each time a tool surpassed us in capability, we adjusted our role. We moved away from raw execution and toward direction, interpretation, judgment, creativity, and care.

Superintelligence, if it emerges, follows this same pattern — just at a different scale.

To understand why, it helps to be clear about what "superintelligence" actually means.

At its core, superintelligence refers to systems that can outperform humans in many cognitive tasks — analysing information, identifying patterns, generating solutions, and optimising outcomes faster and more comprehensively than we can.

It does not automatically mean:

- consciousness
- desire
- values
- moral understanding
- lived experience

It means *capability*, not *purpose*.

And capability alone has never been the source of meaning.

A system may generate better forecasts.
It may propose more efficient plans.
It may identify risks humans miss.

But it does not decide why a goal matters.
It does not define what should be optimised.
It does not choose which trade-offs are acceptable.

Those decisions are not technical.
They are human.

This is where fear often misfires.

We imagine intelligence as a ladder — and fear that once something climbs higher than us, we are pushed off. In reality, intelligence operates more like an ecosystem. Different forms serve different functions. Superiority in one domain does not erase relevance in another.

Being outperformed cognitively is not the same as being replaced existentially.

Just as machines freed humans from physical labour without erasing purpose, advanced cognitive tools can free humans from certain mental burdens without erasing significance. Tasks that require constant calculation, monitoring, or optimisation may increasingly be handled by machines.

What remains — and grows — are the parts of intelligence that do not scale cleanly:

- ethical judgment
- contextual understanding
- emotional resonance
- meaning-making
- responsibility

Superintelligence does not remove the need for these qualities.
It increases the stakes of them.

As systems become more capable, the consequences of decisions become larger. That makes values, oversight, and intent *more* important, not less. Intelligence amplifies outcomes — it does not define which outcomes are worth pursuing.

This is why the conversation around superintelligence often becomes distorted.

Fear frames the question as:
"What happens when we are no longer the smartest?"
A better question is:
"What happens when intelligence becomes abundant?"

When thinking power increases, meaning does not disappear — it becomes the differentiator. Direction matters more than execution. Values matter more than speed. Wisdom matters more than raw capability.

Humanity has always thrived not by being the most powerful force in isolation, but by shaping how power is used.

Superintelligence does not erase the human role.
It repositions it.

What matters is not whether intelligence grows beyond us.
What matters is how we integrate it into systems shaped by ethics, responsibility, and care.

Being outgrown intellectually is not the same as being outgrown existentially.

Human meaning does not vanish when intelligence scales.
It changes context — and context has always been our strength.

The danger is not that intelligence becomes too powerful.
The danger is assuming that power automatically replaces purpose.

It never has.

And it doesn't now.

Chapter 11 — Panic Is Contagious, Calm Is Learned

Fear spreads faster than understanding.

In moments of rapid change, panic behaves like a virus. It moves quickly through headlines, social feeds, conversations, and communities — often detached from evidence, context, or proportional risk. The faster the world appears to change, the more appealing simple, alarming narratives become.

Panic simplifies.
It reduces complexity into threats.
It replaces nuance with certainty.

This is not because people are irrational. It is because uncertainty is uncomfortable. When the future feels unclear, the human mind seeks something solid to hold onto — even if that "certainty" is fear.

Calm, by contrast, does not go viral on its own.

It does not shout.
It does not shock.
It does not demand immediate attention.

Calm requires effort. It must be learned, practiced, and reinforced — especially when everything around us is accelerating.

This imbalance distorts our perception of reality.

Extreme voices dominate attention because fear captures clicks, views, and engagement. Thoughtful analysis moves more slowly. It asks people to pause, reflect, and tolerate uncertainty — and that is harder work. As a result, calm perspectives are often drowned out, not because they are wrong, but because they are quieter.

Caution is mistaken for denial.
Optimism is mislabelled as ignorance.
Balance is dismissed as complacency.

We have seen this pattern repeatedly — not just with technology, but with health scares, economic shifts, political change, and social movements. Panic spreads because it feels urgent. Calm spreads only when people are willing to sit with discomfort long enough to understand what is actually happening.

Artificial intelligence magnifies this dynamic.

The speed of technological change compresses timelines. Developments that once took decades now unfold over years or months. This compression creates the illusion that control is slipping away — even when it is not. When understanding lags behind capability, fear fills the gap.

This is where panic becomes contagious.

One alarming headline is shared.
One dramatic quote is amplified.
One speculative scenario is treated as inevitability.

Before long, a narrative forms — not from evidence, but from repetition.

And repetition feels like truth.

Panic narrows vision.
It pushes thinking into extremes.
It frames the future as something that happens *to* us, rather than something we participate in shaping.

Calm does the opposite.

Calm expands vision.
It allows room for context.
It makes space for questions instead of conclusions.

Calm does not deny risk. It places risk in proportion. It distinguishes between what is possible, what is probable, and what is happening now. It recognises that uncertainty does not equal danger — and that speed does not equal loss of control.

Artificial intelligence does not demand urgency in thought.
It demands clarity.

The faster systems evolve, the more brittle fear-driven decisions become. Policies rushed through panic often overcorrect. Reactions driven by alarm tend to sacrifice long-term stability for short-term relief. History is full of examples where fear produced consequences more damaging than the original threat.

When calm leads, systems remain adaptable.
When fear leads, systems fracture.

This is why calm is not passive.
It is an active discipline.

It requires slowing internal reactions even when the external world accelerates. It requires resisting the pull of certainty when certainty is unwarranted. It requires holding complexity long enough for understanding to form.

The future will not be shaped by those who shout the loudest.
It will be shaped by those who think the clearest.

Calm does not mean disengagement.
It means deliberate participation.

And in moments of rapid change, that may be the most important skill we have.

Chapter 12 — Education Is the Real Bottleneck

Technology rarely advances at the same pace as understanding.

Artificial intelligence is no exception. While tools improve rapidly, the systems that help people understand those tools move far more slowly. This gap — between capability and comprehension — is where confusion, fear, and misinformation take hold.

When people encounter something powerful that they do not understand, imagination rushes in to fill the space. And imagination, when fuelled by uncertainty, tends to lean toward extremes.

This is not a new problem.
It is a familiar one.

We have seen it with electricity, with genetics, with the internet, and with automation. Each time, the technology advanced faster than public understanding. Each time, fear flourished in the absence of clear explanation.

Education is often misunderstood in these moments.

It does not mean turning everyone into an engineer.
It does not mean teaching code to every child.
It does not mean memorising technical details.

Education, in this context, means learning how to *think* about change.

It means understanding what a tool can do — and what it cannot.
It means recognising the difference between capability and consequence.

It means asking better questions rather than jumping to faster conclusions.

When education lags, narratives fill the vacuum.

Simplistic stories spread more easily than nuanced truth. "AI will save us" and "AI will destroy us" are both easier to digest than reality. Complexity is reduced to slogans. Uncertainty is framed as threat. The middle ground — where understanding lives — is crowded out.

This is not a failure of intelligence.
It is a failure of access.

Most people are not resistant to learning.
They are resistant to being overwhelmed.

When information arrives without context, without language that makes sense, and without time to process, it triggers defence rather than curiosity. People disengage, not because they don't care, but because the conversation feels inaccessible.

Fear thrives where people feel excluded.

The moment people are given clear language, grounded examples, and space to ask questions, something changes. Fear softens. Curiosity returns. Agency reappears. The future stops feeling like something happening *to* them and starts feeling like something they can participate in shaping.

This is why education is the real bottleneck — not technology.

We do not lack intelligence.
We lack shared frameworks.

We lack common reference points that allow people from different backgrounds, ages, and professions to understand what is happening without needing specialised expertise. Without those frameworks, people default to emotion, instinct, and social cues — which are easily influenced by panic.

Education does not eliminate risk.
It makes risk legible.

When people understand how systems work, even at a high level, they can distinguish between speculation and reality. They can evaluate claims. They can resist alarmism without dismissing caution.

They can hold complexity without being paralysed by it.
This is how societies adapt successfully.

Not by racing technology forward blindly.
But by bringing understanding along with it.

The most important investment we can make is not in faster systems, larger models, or more automation.

It is in shared understanding.

Because when people understand what is happening, fear loses its grip.

And when fear loosens, better decisions become possible.

Chapter 13 — Ethics Are Not Optional

As artificial intelligence becomes more capable, ethical questions stop being abstract and start becoming operational.

In the early stages of any technology, ethics often feel theoretical — something to be debated later, once the tools are finished. But as systems move from experimentation into everyday use, ethics shift from philosophy into practice. Decisions that once felt distant begin to affect real people, real outcomes, and real lives.

Ethics are often framed as obstacles to progress.
As limitations.
As things that slow innovation down.

In reality, ethics are what allow innovation to *scale*.

Without ethical frameworks, trust erodes. Without trust, systems fragment. And without shared standards, power quietly concentrates in ways that are difficult to see until damage has already occurred.

Every system reflects values — whether intentionally or not.
What is prioritised.
What is ignored.
What is optimised for.
What is constrained.

These are not technical choices.
They are ethical ones.

Choosing speed over safety is an ethical decision.
Choosing efficiency over fairness is an ethical decision.
Choosing profit over accountability is an ethical decision.

The absence of explicit ethics does not create neutrality.
It creates drift.

When no values are clearly defined, systems default to whatever incentives are strongest. And incentives, left unchecked, tend to favour efficiency, scale, and power — not fairness, care, or long-term consequence.

This is not unique to artificial intelligence.

We have seen it in finance, where unchecked optimisation led to systemic risk.

We have seen it in social media, where engagement metrics shaped behaviour in unintended ways.
We have seen it in industrial systems, where efficiency sometimes came at the cost of safety or wellbeing.

Artificial intelligence magnifies these dynamics because it operates at scale and speed. Small design decisions can produce large downstream effects. What once affected hundreds can now affect millions.

That is why ethics are not optional in this moment.
They are foundational.

Clear ethical standards do not slow progress.
They stabilise it.

They give developers, organisations, and societies shared expectations. They create boundaries that reduce uncertainty. They make accountability visible rather than reactive. Without them, responsibility becomes diffuse — and when responsibility is everywhere, it is effectively nowhere.

Ethics do not exist to stop innovation.
They exist to guide it.

They answer questions technology cannot answer on its own:
• Just because we *can*, should we?
• Who benefits?
• Who bears the risk?
• What happens when systems fail?
• Who is accountable when outcomes cause harm?

Artificial intelligence does not remove the need for these questions.
It intensifies them.

As systems become more powerful, the consequences of ethical failure increase. Decisions propagate faster. Errors scale wider. Biases embed deeper. Without deliberate ethical consideration, harm does not announce itself — it accumulates quietly.

The future will not be defined by what technology can do.

It will be defined by what we decide it *should* do — and by whether we are willing to hold ourselves responsible for those decisions.

Ethics are not a brake on progress.
They are the steering wheel.

And without one, even the most advanced systems eventually lose direction.

Chapter 14 — Regulation Is Catching Up (Slowly, but Surely)

Regulation is often framed as a reaction.

Something that arrives late.
Something clumsy.
Something that restricts progress after the damage is already done.

In truth, regulation usually follows understanding.

It lags not because society is careless, but because clarity takes time. Rules cannot be written before we understand what a technology actually does, how it is used, and where its risks truly lie. Artificial intelligence is no different.

Every transformative technology passes through the same early phase.

At first, capabilities evolve faster than rules. Use cases multiply. Boundaries are unclear. The landscape feels unregulated — not because no one cares, but because the questions themselves are still being formed.

As understanding grows, standards begin to emerge.
Expectations take shape.
Limits are discussed, tested, revised, and renegotiated.

This process is rarely neat.
But it is persistent.

The absence of perfect regulation today does not mean the absence of regulation tomorrow.

History is clear on this point.

We regulated medicine once treatments became powerful enough to cause harm if misused.

We regulated aviation once flight became common and stakes became clear.

We regulated finance once systems grew complex enough to threaten stability.

We regulated data once information became valuable enough to require protection.

Each time, regulation followed capability.
Each time, the path was uneven.
Each time, the alternative — doing nothing — proved unacceptable.
Artificial intelligence is following the same arc.

Early tools spark experimentation.
Widespread adoption reveals unintended consequences.
Public awareness grows.
Pressure builds.
Frameworks form.

This is not a failure of governance.
It is how governance adapts.

Regulation is often misunderstood as an attempt to stop progress.
In reality, its purpose is alignment.

It exists to ensure that innovation serves shared values rather than narrow incentives. It sets guardrails, not ceilings. It creates trust, not stagnation. Without regulation, progress accelerates unevenly — benefiting some while exposing others to risk.

Markets understand this intuitively.

As AI becomes more embedded in business, healthcare, education, and public systems, oversight becomes a requirement rather than an obstacle. Consumers demand accountability. Institutions demand reliability. Investors demand predictability. Trust becomes a competitive advantage.

Regulation does not arrive only from governments.
It emerges from many directions at once.

Industry standards.
Professional norms.
Liability frameworks.
Public expectations.
Cultural boundaries.

Together, these forces shape how technology is used long before formal laws are finalised.

This is why the fear that artificial intelligence will remain permanently "unregulated" misunderstands how societies respond to power. Oversight does not appear overnight — but it does appear.

As artificial intelligence becomes woven into daily life, regulation will not be optional.

It will be demanded by citizens who want safety.
By organisations that want clarity.
By markets that want stability.
And by governments tasked with protecting the public interest.

Regulation is not the enemy of innovation.
It is the mechanism that allows innovation to endure.

Progress that cannot be trusted cannot last.
Progress that cannot be governed cannot scale.

Artificial intelligence will be regulated not because it is feared, but because it matters.

And what matters is always shaped — eventually — by shared rules.

Chapter 15 — The Myth of the Sudden Collapse

Apocalyptic stories are appealing because they simplify complexity into a single dramatic moment.

A collapse.
A takeover.
An ending.

They give fear a clear shape. One event. One date. One point of no return.

Reality rarely works that way.

Civilisations do not disappear overnight. Systems bend, strain, adapt, and reform. When disruption occurs, it unfolds unevenly — affecting regions, industries, and individuals at different speeds. Some areas feel change early. Others lag behind. Some benefit. Others struggle. Artificial intelligence will follow the same pattern.

The fear of sudden collapse assumes fragility where there is, in fact, resilience.

Human societies are not delicate structures waiting to shatter. They are layered, redundant, and adaptive systems built over generations. They absorb shocks, reorganise, and continue — not because disruption is painless, but because adaptation is built into how societies function.

We have lived through moments that felt apocalyptic at the time.

World wars that reshaped borders and identities.
Pandemics that disrupted daily life and exposed systemic weaknesses.
Economic crashes that erased wealth and redefined work.
Environmental events that forced migration and reorganisation.

None of these moments were easy.
None of them were evenly experienced.
But none erased humanity's capacity to rebuild and recalibrate.

Each time, new systems emerged alongside old ones. Roles shifted.
Priorities changed. Institutions reformed. Life continued — altered,
but not ended.

Artificial intelligence enters this long pattern of change.

It does not arrive as a singular shockwave that wipes systems away. It
integrates into existing structures gradually, unevenly, and
imperfectly. Some industries feel disruption early. Others adapt
slowly. Some jobs change quickly. Others barely notice at first.

This unevenness is often mistaken for collapse.

When viewed from close proximity, change feels catastrophic. Losses
are immediate. Uncertainty is loud. What is disappearing is easier to
see than what is forming. But step back, and patterns emerge.

Progress does not move as a straight line.
It moves as a series of adjustments.

Old structures strain.
New ones form.
Hybrid systems emerge in between.

This is how transitions work.

The myth of sudden collapse persists because it is emotionally efficient. It turns uncertainty into certainty. It replaces a complex, unfolding process with a single, frightening conclusion. But that clarity is false.

The future shaped by artificial intelligence will not arrive as an ending.

It will arrive as a transition.

Transitions are uncomfortable.
They expose weaknesses.
They demand learning.
They redistribute power and responsibility.

But they are survivable.

Human history is not a story of uninterrupted stability.
It is a story of repeated adjustment.

And each time, humanity has carried forward the same essential capacities:
cooperation,
creativity,

meaning-making,
and adaptation.

Artificial intelligence does not erase those capacities.
It activates them.

The challenge ahead is not surviving a sudden collapse.
It is navigating an extended period of change.

And that is something humans have done before — many times.

Chapter 16 — Choosing How This Story Is Told

Every generation inherits a story about its future.

Sometimes that story is shaped by hope.
Sometimes by fear.
Most often, by a mixture of both.

What matters is not which emotions arise first — fear is a natural response to change. What matters is which emotions we allow to guide our decisions once the initial reaction has passed.

Artificial intelligence will be written into history whether we participate consciously or not. The tools will be built. The systems will be deployed. The changes will unfold. The question is not whether the story will be told — it is who tells it, and from what mindset.

Stories matter because they shape behaviour.

A future framed as inevitable collapse encourages withdrawal. People disengage. Responsibility feels pointless. Why prepare if outcomes are already decided? Why participate if control is assumed to be lost?

Fear-driven narratives frame the future as something that happens *to* us. They position humans as passive observers watching inevitability unfold. In these stories, technology is the protagonist and humanity is reduced to an audience.

Calm narratives do something different.

They restore agency.

They remind us that systems are built, not discovered. That choices are made at every stage — in design, deployment, regulation, and use. That responsibility does not vanish when complexity increases. It concentrates.

This does not require blind optimism.
Blind optimism ignores risk.
Blind fear exaggerates it.

What we need instead is grounded participation.

Participation means staying engaged even when outcomes are uncertain. It means learning enough to ask meaningful questions. It means refusing to outsource all responsibility to institutions, experts, or narratives that benefit from attention rather than understanding.

The story we tell ourselves shapes how we respond long before any law is written or system is deployed.

It shapes how we educate — whether we equip people to think critically or leave them overwhelmed.
It shapes how we regulate — whether we react in panic or build thoughtful guardrails.

It shapes how we innovate — whether we optimise blindly or design with care.

It shapes how we respond — whether we freeze, flee, or participate.

Stories do not just describe the future.

They prepare us for it.

This is why narrative matters so deeply in moments of transition. The story we accept becomes the lens through which every new development is interpreted. It determines whether change feels like an attack or an invitation.

The future is not a script already written.

It is a draft.

And drafts are revised through participation, not prediction.

We are still holding the pen — not because we control every outcome, but because we influence direction through countless small decisions. Through curiosity instead of certainty. Through engagement instead of avoidance. Through responsibility instead of resignation.

Choosing how this story is told does not mean denying uncertainty. It means refusing to surrender agency.

And that choice — made quietly, repeatedly, and collectively — is what shapes the future far more than any single technology ever could.

The future is not a script already written.
It is a draft — and we are still holding the pen.

Chapter 17 — We Are Not Doomed

Every technological age has carried its own version of the same fear:

This time is different.
This time we've gone too far.
This time we won't adapt.

And every time, that belief has been rooted less in evidence than in uncertainty.

Artificial intelligence feels overwhelming because it touches so many aspects of life at once. It reaches into how we work, how we learn, how we create, how we communicate, and how we understand intelligence itself. When multiple foundations shift simultaneously, fear can feel not only natural, but rational.

But fear is not foresight.
Fear reacts to uncertainty.
Foresight engages with it.

Throughout this book, we have examined the sources of anxiety surrounding artificial intelligence — not to dismiss them, but to understand them. We have explored why fear spreads, why panic feels contagious, why control feels threatened, and why the future can feel as though it is narrowing rather than opening.

What emerges, again and again, is a consistent pattern:

The danger is not the technology.
The danger is misunderstanding it.
The danger is disengaging from it.

The danger is surrendering responsibility before it is actually taken from us.

We are not standing at the edge of an ending.

We are standing at the beginning of a reorganisation.

Reorganisations are rarely comfortable. They are uneven, imperfect, and sometimes disorienting. Old structures strain. Familiar roles shift. Assumptions are challenged. But reorganisations are not collapses — they are transitions.
And transitions are survivable.

Humanity's greatest strength has never been control over its environment. We have never fully controlled nature, economies, or change itself. What we have always possessed is something more flexible and more powerful:

The ability to adapt meaning, values, and cooperation as circumstances change.

This is what carried us through previous technological revolutions.
This is what allowed societies to rebuild after disruption.
This is what has repeatedly transformed fear into progress.

Artificial intelligence does not remove that strength.

It tests whether we are willing to use it again.

The tools are changing.
The pace is accelerating.
The questions are becoming more complex.

But the core human capacities that matter most remain unchanged.

Judgment.
Ethics.
Empathy.
Creativity.
Responsibility.
Meaning.

These are not things that scale automatically.
They require participation.

We do not need blind optimism.
We do not need resignation.
We do not need panic.

We need engagement.

We need people willing to learn just enough to ask better questions.
We need institutions willing to prioritise long-term trust over short-term speed.

We need conversations grounded in clarity rather than alarm.

We need to remember that intelligence alone does not determine the future — values do.
Artificial intelligence will shape the world ahead.
So will the choices we make around it.

Those choices will not be made in one dramatic moment. They will be made gradually, through design decisions, policies, cultural norms, and everyday use. Through what we reward. Through what we tolerate. Through what we question.

This is not a story about machines replacing humanity.

It is a story about whether humanity remains present as its tools grow more powerful.

We are not doomed.

We are being asked to grow — not by becoming something else, but by becoming more intentional about who we already are.

Growth has never been easy.
It has never been fast.
And it has never been free of uncertainty.

But growth has always been possible.

And this moment is no different.

Final Note — After the Noise

This book ends quietly on purpose.

Much of the conversation around artificial intelligence is loud. Urgent. Absolute. It demands certainty in a moment that doesn't yet offer it. But clarity rarely arrives through noise. It arrives through reflection.

Artificial intelligence will continue to advance. That is not a belief — it is already happening. What remains undecided is not whether the technology will change the world, but how we will change alongside it.

Fear is an understandable response to rapid transformation. But fear is not a strategy. It narrows perspective, accelerates division, and convinces us that complexity must have a single villain or a single outcome. History suggests otherwise. Progress has never been clean, but neither has it ever been final.

What matters now is not prediction, but posture.
Curiosity over certainty.
Participation over panic.
Responsibility over resignation.

We are not spectators to this moment. We are contributors to it. The systems being built today will reflect the values we choose to apply — or fail to apply — right now. That responsibility does not belong only to engineers, companies, or governments. It belongs to societies, communities, and individuals willing to engage thoughtfully rather than react reflexively.

If this book has done anything, I hope it has created space — space to slow the conversation down, to question the extremes, and to remember that intelligence, artificial or otherwise, does not determine meaning on its own. Humans still do.

The future is not waiting to happen to us.

It is being shaped — quietly, continuously — by the choices we make, the narratives we repeat, and the mindset we carry forward.

We are not doomed.

We are deciding.

— Michael Besnard

www.ingramcontent.com/pod-product-compliance
Lightning Source LLC
Chambersburg PA
CBHW032359280326
41935CB00008B/630